Chemical Engineering & the Law Forum 2022

Held at the 2022 AIChE Spring Meeting and 18th Global Congress on Process Safety

San Antonio, Texas, USA
10 – 14 April 2022

ISBN: 978-1-7138-6832-3

Printed from e-media with permission by:

Curran Associates, Inc.
57 Morehouse Lane
Red Hook, NY 12571

Some format issues inherent in the e-media version may also appear in this print version.

Printed with permission by Curran Associates, Inc. (2023)

For permission requests, please contact AIChE
at the address below.

AIChE
120 Wall Street, FL 23
New York, NY 10005-4020

Phone: (800) 242-4363
Fax: (203) 775-5177

www.aiche.org

Additional copies of this publication are available from:

Curran Associates, Inc.
57 Morehouse Lane
Red Hook, NY 12571 USA
Phone: 845-758-0400
Fax: 845-758-2633
Email: curran@proceedings.com
Web: www.proceedings.com

TABLE OF CONTENTS

Chemical Processes and Trade Secret Protection 1
Lauren Dowty, Paul Townsend, Andrew Renison, Charles Collins-Chase

Bars, Experiments and Threats: Lessons for Engineers Who Deal with Inventions from Energy Heating, LLC V. Heat on-The-Fly, LLC 2
Gerard Reinhardt

Author Index

639558 Chemical Processes and Trade Secret Protection

Abstract:

Chemical processes are the backbone to numerous products making their way into the market in many industries. Despite the large chemical processes industry, intellectual property protection for such processes is less than clear. Companies may use patents to protect certain chemical processes; however, patent protection requires public disclosure of the process. If a company does not wish to publicly disclose its chemical process, it is left without the availability of patent protection. How then do companies keep chemical processes protected?This presentation will provide an overview of the basics of IP, specifics on trade secret law, and important trade secrets considerations for companies wishing to protect chemical processes. Speakers will also highlight several case studies addressing chemical processes trade secrets and lessons learned from those cases.

Bars, Experiments and Threats:
Lessons for Engineers Who Deal with Inventions from
Energy Heating, LLC v. Heat On-The-Fly, LLC

by

Gerard E. Reinhardt
Reinhardt IP, PA
ReinhardtIP.com
Gerard@ReinhardtIP.com

Cite as "Gerard E. Reinhardt, 'Bars, Experiments and Threats:
Lessons for Engineers Who Deal with Inventions from
Energy Heating, LLC v. Heat On-The-Fly, LLC'".

Introduction

The U.S. Court of Appeals for the Federal Circuit[i] ("the Federal Circuit") issued a decision in *Energy Heating, LLC v. Heat On-The-Fly, LLC*[ii] on October 14, 2021. This is a patent case that illustrates the application of several key aspects of patent law, and offers cautionary lessons to any practitioners whose clients may include inventors, patentees or accused patent infringers, as well as to general litigators.

Background

Heat-on-the Fly, LLC ("HOTF") is the owner of U.S. Patent No. 8,171,993[iii] ("the '993"). Energy Heating LLC and Rocky Mountain Oilfield Services, LLC (collectively "Energy") provide water heating services for water used during the hydraulic fracturing process ("fracking.")

Ransom Mark Hefley ("Hefley") was a founder and part owner of HOTF, and is the sole inventor named on the '993. The '993 is directed to methods, systems and an apparatus for heating water on demand or inline during the fracking process instead of using preheated water in large standing tanks. Hefley filed with the U.S. Patent & Trademark Office ("the USPTO") the earliest provisional application[iv] to which the '993 claims priority on September 18, 2009. Prior to the critical date of the '993 (one year before the earliest priority date), HOTF sold sixty-one heat-on-the-fly systems for a total of $1.8 million. During prosecution of the nonprovisional patent, HOTF failed to disclose the prior sales to the USPTO, in violation of its duty of disclosure[v]. The USPTO granted the '993 on May 8, 2012.

Litigation

Energy filed a petition under the Declaratory Judgment Act[vi] in January 2013, seeking a declaratory judgment that the '993 patent was, *inter alia*, unenforceable for inequitable conduct[vii]. Energy's theory of inequitable conduct was based on HOTF's intentional failure to disclose the pre- critical date sales to the USPTO in violation of its duty of disclosure, and that but for that failure, the USPTO would not have granted the '993 due to invalidity under the statutory bars of 35 USC 102.[viii] In response, HOTF counterclaimed for infringement of the '993, inducement to infringe, and contributory infringement against Energy.

The district court held a bench trial on inequitable conduct during which HOTF contended that the pre- critical date sales did not trigger the duty to disclose them to the USPTO during prosecution because they constituted experimental uses[ix], and that on that basis there was no inequitable conduct. The court found that the pre- critical date uses of the invention were neither experimental nor conducted in secret, but rather were pursuant to sales for revenue.[x] Based on the foregoing, the court granted a declaratory judgment against HOTF on the issue of inequitable conduct,[xi] but denied Energy's motion for a finding of exceptionality[xii] and an award of attorneys' fees and costs under *§ 285* without further explanation.[xiii] The court also found $750,000 in actual damages based on a state unfair trade practices statute based on HOTF's contention that the patent was valid and enforceable made during telephone conversation threatening an Energy customer with patent litigation.

On appeal, the Federal Circuit affirmed the judgment that the '993 patent is unenforceable due to inequitable conduct,[xiv] but vacated the district court's denial of attorneys' fees under *§ 285* for want of a detailed justification, and remanded the issue to the district court for reconsideration.[xv]

On remand, the district court reversed its position, finding that the case was exceptional and awarding attorneys' fees and costs of $5,312,307.[xvi] The holding of exceptionality was based on the findings that "this case stands out from others with respect to the substantive strength of HOTF's litigation position" and that "HOTF litigated the case in an unreasonable manner by persisting in its positions." The district court also cited the very large number of pre- critical date sales that HOTF failed to disclose to the USPTO (61), and the substantial revenue derived from those sales ($1.8 million, which undermined HOTF's contention of experimental use) for its finding of exceptionality.

Holding

On appeal, the Federal Circuit affirmed, stating that a finding of litigation misconduct is not necessary to find a case exceptional. "[A]s a general matter, many forms of misconduct can support a district court's exceptional case finding ..." The Federal Circuit found that the district court properly considered the totality of the circumstances, including the manner of HOTF's litigation, finding that "HOTF litigated the case in an unreasonable manner by persisting in its positions [the contention that the 61 pre- critical date sales were all experimental uses, which did not trigger the duty of disclosure]."[xvii] This result is striking in the absence of any of the usual allegations of litigation misconduct, e.g., discovery abuse, dilatory or harassing tactics, subornation of perjury.

In summary, HOTF wound up not only losing its patent, but also paying out a total of about $6.25 million in attorneys' fees, costs and damages (as well as their own attorneys' fees and costs) by attempting to enforce a patent based on a flawed legal contention. Contrast this with the $1.8 million in revenue that HOTF gained from the 61 sales of the invention from 2006 to 2012.

Practice Tips

This case provides a variety of abject lessons for inventors, those involved in licensing patents, patentees contemplating enforcement, and those accused patent infringement, including the following:

1. **Inventors should file a patent application (provisional or nonprovisional) with the USPTO before the first offer for sale, sale or public use. Failing that, file within one year of the first offer, sale or public use.** If HOTF had conformed to this simple rule, the results of this case would have been very different.

2. **To preserve the experimental use exception to the statutory bar defenses, inventors should conduct any pre-filing experiments in-house or with an independent contractor or other third party under a duty of nondisclosure and confidentiality, and maintain lab notebooks documenting the conditions and results of the use.** HOTF failed to comply with any of these indicia of experimental use, and thus was subject to the statutory bar and inequitable conduct defenses.

3. **Prior to filing, disclose any pre-filing offers for sale, sales or public uses, as well as all relevant prior art under 35 USC §102, to the attorney prosecuting the patent application.** Upon receipt of such information, he/she will evaluate it, and if required, should timely disclose it to the USPTO. Had HOTF conformed to this requirement, the '993 would likely not have been granted – on the whole a better outcome for all involved in this case.

4. **Consult with patent counsel before communicating any threat of litigation to a potential accused infringer. Recipients of such threats should also immediately consult patent counsel.** HOTF made a verbal threat of litigation, which provided Energy with the basis to bring a Declaratory Judgment action, and thus gain the advantages of a plaintiff regarding choice of venue, timing and other procedural matters. The allegation that HOTF also verbally communicated that the '993 was valid and enforceable could have been the basis of an award of treble damages had Energy pleaded appropriately.

5. **Consult with a patent attorney to conduct a thorough scrub down of any patent under consideration for enforcement. Decline to enforce any**

patent that bears significant risks of inequitable conduct. If a patentee wants to initiate a dialogue with a potential infringer in the hope of avoiding litigation, his/her patent attorney should draft a Cease and Desist letter worded to minimize the chances of success of a declaratory judgment action by the recipient. Be prepared to file a Complaint in short order thereafter if a satisfactory result is not obtained. Be mindful that under FRCP 11[xviii], "[b]y presenting to the court a pleading, written motion, or other paper... an attorney... certifies that to the best of the person's knowledge, information, and belief, formed after an inquiry reasonable under the circumstances... (2) the claims, defenses, and other legal contentions are warranted by existing law or by a nonfrivolous argument for extending, modifying, or reversing existing law or for establishing new law..."

6. **Reconsider maintaining a failing position in litigation.** Had HOTF withdrawn its infringement claims as discovery showed the strength of Energy's inequitable conduct defense, it likely would have avoided the holding of exceptionality and the award of $5.5 million in attorney's fees and costs. Indeed, Rule 11 may encompass a duty to withdraw a baseless pleading.[xix]

REINHARDT IP, PA
INTELLECTUAL PROPERTY DELIVERED SMARTLY

Gerard E. Reinhardt
Gerard@ReinhardtIP.com
Ph: 516-672-3988
915 SE Krueger Parkway
Stuart, FL 34996
ReinhardtIP.com

In over twenty years of patent law experience, Mr. Reinhardt has developed a sophisticated practice focused on the creation of commercial value, including:

- **Drafting & prosecuting U.S and foreign patent applications**;
- **Patent licensing** transactions;
- **Due diligence** assessments;
- **Patent portfolio management,** including preparation of concise summaries of key features of each asset for ease of evaluation by prospective licensors/purchasers;

- **Patent brokerage;**
- **Opinions** regarding patentability, freedom-to-operate, infringement and validity;
- Enforcement and defense in **patent disputes**;
- Corresponding services regarding **trademarks, trade secrets, copyrights** and **domain names**;
- Sales of businesses; and,
- Legal matters associated with the above (e.g., contracts, tortious interference, unfair competition, privacy, matters involving engineering or technology).

Mr. Reinhardt holds BS and MS degrees in Chemical Engineering, and has twenty-five years of industry experience in hydrocarbon processing, fluoropolymers, pharmaceuticals, avionics and jet engines. He is licensed to practice law in Texas, New York, Florida and the District of Columbia, as well as before the U.S. Patent & Trademark Office.

---

[i] The Federal Circuit (a court of twelve judges that sits in panels of three, and is based in Washington, D.C.) has sole jurisdiction over appeals of patent cases from district courts under *28 USC §1295(a)(1)*. A writ of certiorari may be properly filed upon issuance of a final judgment from the Federal Circuit. As a specialized court of appeals, and one from which certiorari is relatively infrequently taken, the Federal Circuit jurisprudence is widely considered to be authoritative in matters of patent law.

[ii] *2021 U.S. App. Lexis 30690, _ F 4th _ 2021 WL4782688.*

[iii] The '993 includes 99 claims. As an example, Claim 76 recites the following:
An oil well hydraulic fracturing system, comprising: a) a transportable heating apparatus that heats water to a temperature of at least 40 degrees F. (4.4 degrees C.); b) a source of cool or cold water at about ambient temperature; c) a mixer having an inlet, an outlet and a mixer bore that extends between the inlet and the outlet; d) a first lateral fitting on the mixer that enables heated water to enter the mixer bore; e) a second lateral fitting on the mixer that enables removal of water from the mixer bore upstream of the first lateral fitting and wherein at least one of the lateral fittings has a wall portion that extends into the mixer bore; f) a first flowline that transmits heated water between the heater and the first lateral fitting; g) a second flowline that transmits water between the second lateral fitting and the heater, the second flowline being upstream of the first lateral fitting; and h) a mixing tank that is receptive of a flow of a mix of cool or cold and heated water from the bore of the mixer, said tank enabling a proppant to be mixed with the mix of cool or cold and heated water that is discharged from the mixer outlet.

[iv] U.S. Provisional No. 61/276,950.

[v] The filing of a patent application with the USPTO triggers a duty of disclosure and a duty of candor to the USPTO. "Each individual associated with the filing and prosecution of a patent application has a duty of candor and good faith in dealing with the Office, which includes a duty to disclose to the Office all information known to that individual to be material to patentability as defined in this section... [N]o patent will be granted on an application in connection with which fraud on the Office was practiced or attempted or the duty of disclosure was violated through bad faith or intentional misconduct.." *37 CFR 1.56, Duty to disclose information material to patentability.* The duties endure from the filing date until the patent application is granted.

[vi] Under the *Declaratory Judgment Act, 28 USC § 2201*, "[i]n a case of actual controversy," an interested party may file a pleading in US district court to obtain a declaration of rights and legal relations "whether or not such further relief is or could be sought. Any such declaration shall have the force and effect of a trial judgment or decree and shall be reviewable as such." A threat by a patentee of an enforcement action against an accused infringer has commonly been accepted as providing the basis for finding a "case or controversy," and thus declaratory judgment jurisdiction, under the statute.

[vii] Inequitable conduct is an equitable defense to a claim of patent infringement in which the accused patentee is alleged to have failed to meet the duty of disclosure or the duty of candor during prosecution of the patent application in question before the USPTO. To prevail on inequitable conduct based on a failure to observe the duty of disclosure, the accused infringer must prove by clear and convincing evidence that the applicant knew of the reference or prior commercial sale, knew that it was material, and made a deliberate decision to withhold it. *Therasense, Inc. v. Becton, Dickinson & Co., 649 F.3d 1276, 1290 (Fed. Cir. 2011) (en banc).* A judicial finding of inequitable conduct results in the patent being held unenforceable.

[viii] A patent is invalid under the on-sale bar of *35 USC 102(b)* if, before the critical date, the invention was both (1) the subject of a commercial sale or offer for sale and (2) ready for patenting. *Pfaff v. Wells Elecs., Inc., 525 U.S. 55, 67, 119 S. Ct. 304, 142 L. Ed. 2d 261 (1998); see also The Medicines Co. v. Hospira, Inc., 881 F.3d 1347, 1351 (Fed. Cir. 2018).* The critical date of a patent for purposes of applying a statutory bar is that date that is one year before the earliest claimed priority date.

[ix] Experimental use is an exception to the statutory bars of *35 102(b)*. If a prior commercial sale was a bona fide experiment to (1) test the claimed features or (2) determine if the invention would work for its intended use, the sale will not serve as a bar. *Clock Spring, L.P. v. Wrapmaster, Inc., 560 F.3d 1317, 1327 (Fed. Cir. 2009).*

The well established indicia of experimental usage are set forth in *Allen Engineering Corp. v. Bartell Industries, Inc., 299 F.3d 1336 (Fed. Cir. 2002)*. These objective factors include (1) the necessity for public testing, (2) the amount of control over the experiment retained by the inventor, (3) the nature of the invention, (4) the length of the test period, (5) whether payment was made, (6) whether there was a secrecy obligation, (7) whether records of the experiment were kept, (8) who conducted the experiment, (9) the degree of commercial exploitation during testing, (10) whether the invention reasonably requires evaluation under actual conditions of use, (11) whether testing was systematically performed, (12) whether the inventor continually monitored the invention during testing, and (13) the nature of contacts made with potential customers. *Id. at 1353* (citing *EZ Dock v. Schafer Sys., Inc., 276 F.3d 1347, 1357 (Fed. Cir. 2002)* (Linn, J., concurring). The district court observed that the experimentation must be related to the limitations of a claim in the as-yet-to-be-filed patent application, and still more prospective granted patent. *Energy Heating 2016 US Dist. Lexus 193994; WL 10837799.*

[x] The court found that Hefley made no attempt to enter into confidentiality agreements with others involved in the jobs in which he used his system for heating water on the fly; nor did he make any attempt to hide the system he was using to heat water. Furthermore, Hefley's claim that the prior uses were experimental was belied by his conduct in conducting the heating. Hefley kept no notebooks, drawings, plans or explanations of the outcomes. He expressed no preliminary hypotheses prior to the heating and recorded no conclusions confirming or rejecting the hypotheses. The absence of contemporaneous records, explanations, or confirmation by uninterested parties demonstrated that the prior uses were not experimentation. Hefley presented no evidence to satisfactorily explain or justify his knowing failure to disclose the prior uses to the United States Patent Office.

[xi] *Energy Heating, LLC v. Heat On-The-Fly, LLC, No. 4:13-cv-10, 2016 U.S. Dist. LEXIS 193994, 2016 WL 10837799, at *4 (D.N.D. Jan. 14, 2016)* (*"Inequitable Conduct Op."*).

[xii] Under 35 USC *§ 285*, a "court in exceptional cases may award reasonable attorney fees to the prevailing party." An "exceptional" case under *§ 285* is "one that stands out from others with respect to the substantive strength of a party's litigating position (considering both the governing law and the facts of the case) or the unreasonable manner in which the case was litigated." *Octane Fitness, LLC v. ICON Health & Fitness, Inc., 572 U.S. 545, 554, 134 S. Ct. 1749, 188 L. Ed. 2d 816 (2014)*. The party seeking fees must prove that the case is exceptional by a preponderance of the evidence, and the district court makes the exceptional-case determination on a case-by-case basis considering the totality of the circumstances. *Id. at 554, 557-58*. Prevailing on a claim of inequitable conduct often makes a case exceptional. *Therasense, at 1289.*

[xiii] *Energy Heating, LLC v. Heat On-The-Fly, LLC, No. 4:13-cv-10, 2016 U.S. Dist. LEXIS 194012, 2016 WL 10837794 (D.N.D. Mar. 16, 2016)* (*"Attorneys' Fees Op."*).

[xiv] *Energy Heating, 889 F.3d at 1296.*

[xv] *Id. at 1307-08.*

[xvi] *Energy Heating v. Co. Heat, 2020 U.S. Dist. LEXIS 255551, 2020 WL 9848847 (D.N.D., June 15, 2020).*

[xvii] *2021 U.S. App. Lexis 30690, _ F 4th _ 2021 WL4782688.*

[xviii] *Fed. R. Civ. P. 11.*

[xix] See, e.g., Julia K. Cowles, Rule 11 of the Federal Rules of Civil Procedure and the Duty to Withdraw a Baseless Pleading, 56 Fordham L. Rev. 697 (1988).

AUTHOR INDEX

Collins-Chase, Charles 1
Dowty, Lauren 1
Reinhardt, Gerard 2
Renison, Andrew 1
Townsend, Paul 1

AIChE
120 Wall Street, FL 23
New York, NY 10005-4020

ISBN 978-1-7138-6832-3

CURRAN ASSOCIATES INC.
proceedings.com